So you want to be a beekeeper?
An Introduction to Beekeeping

J R Slade

NORTHERN BEE BOOKS

So you want to be a beekeeper?
An Introduction to Beekeeping

© J R Slade

All rights reserved. No part of this publication may be reproduced, stored in a retrieval system, transmitted in any form or by any means electronic, mechanical, including photocopying, recording or otherwise without prior consent of the copyright holders.

ISBN 978-1-908904-54-6

Published by Northern Bee Books, 2014
Scout Bottom Farm
Mytholmroyd
Hebden Bridge HX7 5JS (UK)

Design and Artwork, D&P Design and Print
Printed by Lightning Source UK

So you want to be a beekeeper?
An Introduction to Beekeeping

J R Slade

This book belongs to:

Spaces in the text have been left for the addition of notes and photographs.

INTRODUCTION

This is not a textbook in the ordinary sense of the word. It is an introduction to beekeeping and is intended to act as a memory aid for use with children and novices. Its object is to describe how the bees live and how the beekeeper makes use of them.

With any course on beekeeping that a person joins there is the difficulty of formulating a book that fit either a bee calendar or any learning programme; for this reason there is not an exact point by point correlation between any course and order in which subjects are covered in this booklet. Additionally this booklet is designed to encourage people to become beekeepers. The intention is to familiarize new beekeepers with information whilst at the same time instilling confidence in such a way that whatever a person's background they become proficient beekeepers. In all matters relating to beekeeping an understanding of bees is paramount and a developed empathy with bees is essential. Misunderstanding bees and treating them badly is not acceptable. In order to become a competent beekeeper there is much to be learnt about how bees live and how they are handled. There are however, things to be learnt that are not in books, and some of these you can only learn by handling, helping others to keep or keeping bees yourself. If you can spend time just watching bees at work, you will discover for yourself things that no one else can tell you. This book will help you to understand what you see.

HOW TO START BEE KEEPING

One of the best and easiest ways to start beekeeping is to join a local bee group such as a BBKA teaching apiary. Any person who knows very little or nothing about bees should watch the handling of hives and stocks. It is a good thing to find out in this way if you like bees before starting on your own. Joining a bee group and attending a course on beekeeping and getting your hands inside a hive is the only way to ascertain if beekeeping is for you

Start in a small way, either with one established stock on combs, or with a nucleus. An established stock will start collecting honey almost at once, and with proper management and in a good season should produce a surplus of honey. A nucleus, on the other hand, has to grow into a full stock, and this nursery work will keep the bees busy for many weeks before they start storing honey. In fact, they may give no surplus honey the first season. During this time however, one can become used to handling the bees.

A third way of starting is to obtain a swarm. The swarm is placed in a hive that has been cleaned out and fitted with a brood-chamber containing frames and wax foundation. They can be fed with thin syrup for a few days, after which a super can be put on if the weather is kind. Swarms can weigh up to 8 lbs., and there are about 5,000 bees to a lb. Swarm weights vary with the time of year. An early swarm could produce a surplus of honey in its first season.

So you want to be a beekeeper? - *An Introduction to Beekeeping*

The important points to remember are these :

1. Start in a small way and increase the number of hives gradually as your confidence grows.
2. Always keep healthy, strong stocks.
3. Do not handle the bees more than is necessary, and never in cold, wet, thundery, or windy weather, or when the shade temperature is below 16° C (60°) Fahrenheit.
4. Always keep cool, calm, and collected whatever happens. When handling bees, work smoothly and gently. Never be jerky in your movements, for the bees do not like it. Do not try to brush them away; waving your arms about will only make them angry. If a bee crawls up your sleeve, hold your hand in the air, for bees always crawl upwards. If you are wearing long trousers it is a good idea to fasten at the bottom with a pair of cycle clips, tuck them into your socks or wear wellingtons. Before opening a hive make up your mind what you are going to do, have all the necessary appliances handy, and do not leave the hive until the work is completed.
6. Read what other people have written about bees and bee-keeping. Collect facts and opinions, but keep the two distinct.

WHERE TO SITE BEES

The apiary should be in as secluded and sheltered a position as possible. The bees should have a clear flight from the hives, and if for any reason, the hives have to face a path or regularly used area, the bees should be made to fly high by erecting a screen of small mesh wire-netting, by planting a hedge, or by growing a row of tall plants (beans or peas, perhaps) a few feet from the entrances to the hives..

To make the best use of your bees you must have them close at hand, remember that beekeeping can be hard work. Hives sited away from easy access makes getting equipment to the apiary and honey etc from the apiary difficult if every thing has to be carried by hand.

The keeping of hive records opens up a wide field for investigation, a field that is all too neglected. Each hive should have its record card upon which are entered particulars of the stock, the conditions inside the hive, and the work done with the colony throughout the year. This would include a note on how and when the stock was obtained, the number of Queen cells found and destroyed, the introduction of any new Queen, the amount and quality of honey removed, etc. Weather records can be correlated with the activity of the bees, the honey flow, and the kinds of flowers worked. A scale hive (i.e. a hive standing on a spring balance) provides additional detailed information. Accounts will include purchases and sales and a profit and loss account.

Bees, unlike other livestock, do not require daily attention. There are of course, times when immediate attention must be given, swarming for example, but the fact that a daily routine is not involved makes bee keeping an ideal pastime.

Affiliation to an association such as a BBKA branch apiary generally provides insurance in the unlikely event of your bees having to be destroyed because of foul brood.

So you want to be a beekeeper? - *An Introduction to Beekeeping*

Identifying the Queen
The queen is larger than the worker bees but not greatly so.

1. Note the difference in length and width of body; in the shape of head, thorax and abdomen; and in the proportion of length of wing to length of body in worker bees.

2. The Queen's body is longer than the Workers' bodies, and her short wings are neatly folded on her back.

3. The queen also has a large hairless (bald) thorax.

4. When queens are marked, a dot of coloured permanent ink is applied to the bald area.

5. As a form of swarm control most (not all) beekeepers clip one wing pair and reduce their length by about 1/3. This is not injurious to the queen in her normal duties within the hive but does prevent her from flying.

Top bar hive being examined by a bee inspector

Inspection of your bees by an inspector is not compulsory but there is a legal requirement to report both European and American foul brood. If you as a new beekeeper do not feel competent in the recognition of either then an inspection by the local bee inspector is a good idea. The normal inspection period is every other year.

TYPES OF HIVE

Bee hives fall into two basic groups. The first group is what might be described as commercial hives. They are larger and therefore heavier and require greater physical effort in their use and manipulation. The other group of hives are those more suited to the amateur and beginner bee-keepers.

Of the hives most suitable for the beginner; the National hive is the most popular, the second is the WBC. However, a traditional African type of hive is becoming popular because of it's suitability for the lady beekeeper and older beekeeper as there is not so much heavy lifting. This type of hive is called a Top Bar Hive. Historically there is also the bee skep which is not really in use now for keeping bees but is extensively used for taking swarms.

The Skep (New)

Skeps are still used for beekeeping in some parts of Europe but generally the British climate is not suitable. When they are used, two skeps are placed one upon another. The lower one is for the brood and the upper for honey storage. The skep shown above would normally be used to collect a swarm and housing it until a hive became available.

An inverted skep showing comb that has been created naturally

When double skeps are in use, a smaller one stands upon a larger one. The bees enter through the little doorway (as shown on the new one) near the bottom of the larger skep.

When this skep has been filled with comb, the bees go up through a hole in the top of the larger skep into the smaller. Here they build more comb and store the honey. As with all hives the storage of honey is separated from the brood by the use of a queen excluder

A WBC Hive

W.B.C. are the initials of the inventor of this type of hive William Broughton Carr.

WBC hives are double skinned, that means that what you see from the outside is a series of outer nesting boxes called risers (hive in photo above has 4) that go around an inner hive where the bees are.

When looking through the bees it is necessary to remove the roof and the outer boxes before you can start. This type of hive is generally only used by beekeepers that only want 1 or 2 hives and also want the decorative look of the hive in their garden.

Top bar hive being examined

Section through a top bar hive. Note the centre support peg for the comb

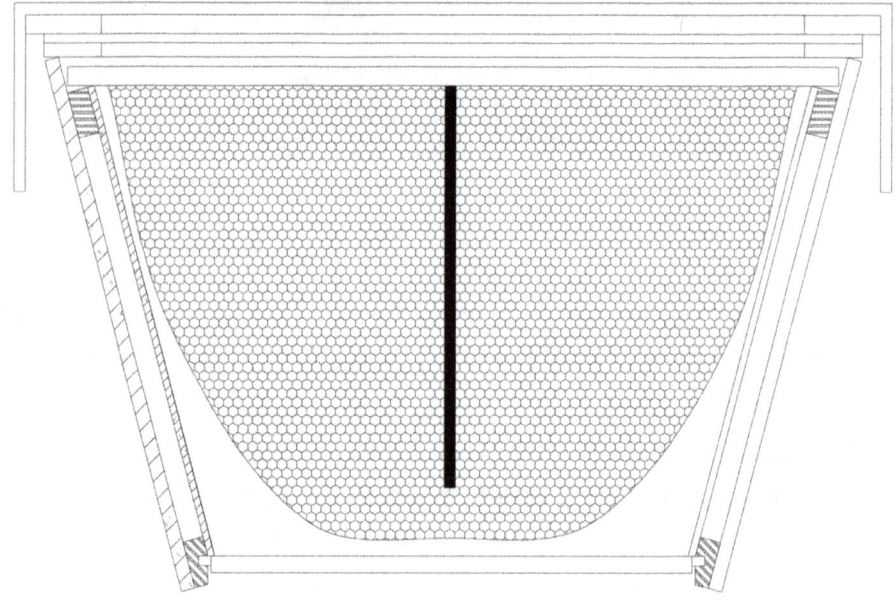

National hive with pitched roof

So you want to be a beekeeper? - *An Introduction to Beekeeping*

Section through a typical single walled hive such as a National

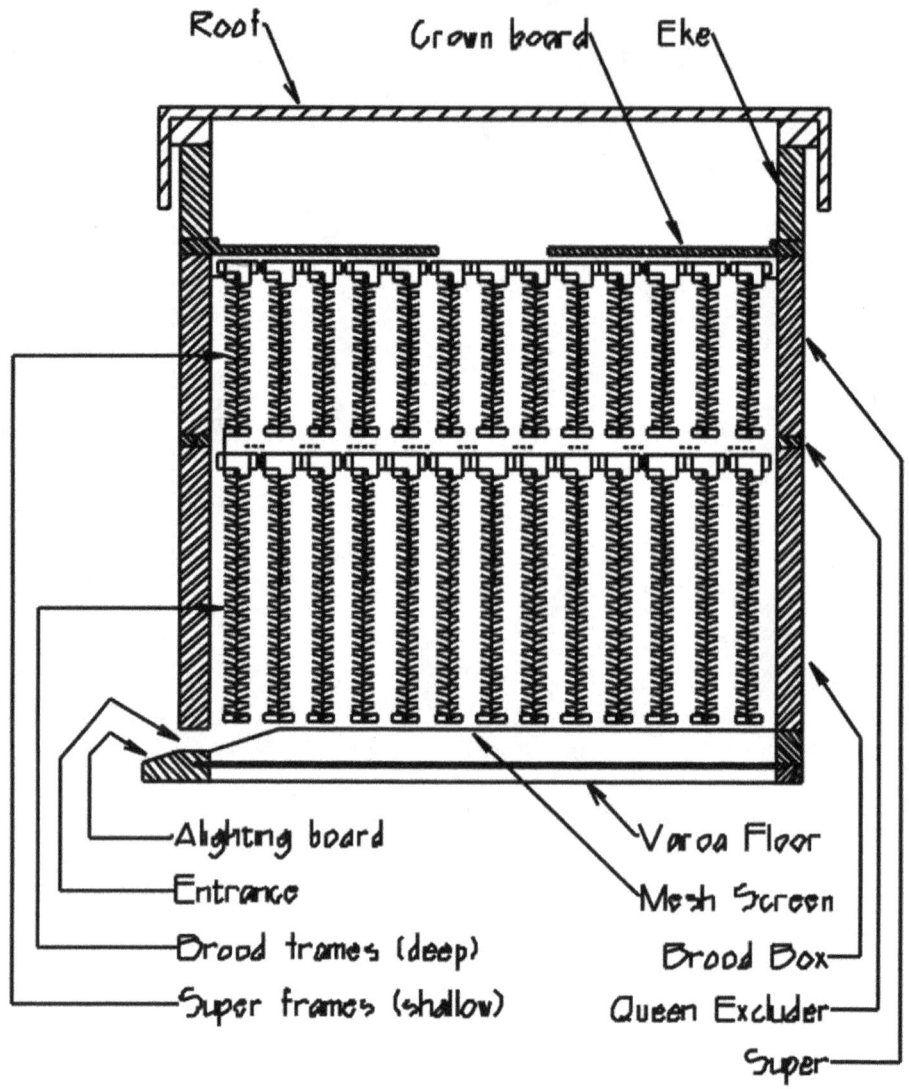

FULLER EXPLANATION OF THE MAIN COMPONENTS OF A SINGLE WALLED HIVE.

Alighting-board, A protruding board upon which the bees land when they arrive at the hive; and from which they take off when leaving.

Entrance to the hive.

Brood frame in brood-chamber. This is the part reserved for breeding, i.e., the laying of the eggs and the rearing of the grubs.

Super, Shallow frame box. In this the bees store their surplus honey.

Sections, Not shown but are also used for storing honey. These are provided only if honey " in the comb " is wanted.

Roof The most important part as it makes the hive weather proof. Its dimensions should be such that water running off does not run down the side of the hive.

Varroa Floor. Most hives are infested with a parasitic mite called varroa. Use of a mesh screen that makes the bottom of the hive bee proof permits debris from the bees to fall through the mesh onto a removable panel. This debris will include the bodies of varroa that have either died or been killed by the bees. The counting of the dead varroa gives an accurate indication as to the degree of infestation in a hive with varroa.

Eke. Ekes are used primarily to give a work space within the hive for the beekeeper. The depth of the eke (top to bottom) can vary according to what the beekeeper wants or needs. It may be used to give space for a feeder or for the addition of insulation material during the winter.

Crown board. The crown board is the barrier that limits the movement of bees upwards. The crown board has a hole in it in the shape of a

So you want to be a beekeeper? *- An Introduction to Beekeeping*

lozenge. The hole permits the bees to ventilate the hive, to access a feeder if in use or to locate a devise (porter bee escape) for removing bees from parts of the hive when the beekeeper requires. Crown boards are also used in conjunction with a Porter bee escape for removing honey.

Queen excluder. This is a slotted barrier that permits only worker bees to pass through, thus preventing the queen from laying eggs in the super where honey is to be stored.

Bees dancing

The bee dance (waggle dance) is performed by returning foraging bees to pass information to other foraging bees about the direction and distance of a nectar source.

Bee fanning

Note. The Nasonov gland is exposed at the top of its abdomen.
The Nasonov gland produces pheromones that are distinctive to the hive where the bee lives. When a hive is disturbed bees can be seen fanning and sending out a "homing scent". This is because a hive, contains a lot of nurse bees that can fly but have not left the hive and have not orientated themselves to the location of the hive. The emission of the pheromones acts as a location beacon so that the nurse bees know which is their hive. The emission of pheromones, also take place during swarming. Once air born, the bees within the swarm are aware of the others and thus can remain as a coherent colony.

THE OCCUPANTS OF THE HIVE

The occupants of a hive of bees are the Workers, the Queen, and the Drones. At all times of the year there are also eggs and the larvae that hatch from these eggs.

The grubs develop later into young bees. The eggs and larvae are generally referred to as *brood*.

The number of bees in a hive varies according to the time of year. In the winter, when they are quiet, there may be about 10,000, whilst in the summer, when they are very busy, there may be four or five times as many. Most of these will be Worker bees.

The Workers. These do all the work of the hive. In the summer they work so hard that they only live for about six weeks. After that time their wings become so torn that they cannot fly, and they crawl away from the hive to die. At other times of the year, when they do less work, they live for about six months.

They start work almost as soon as they have developed into adult bees. One of their first duties is to clean out the cells in which the Queen will lay eggs. Inside the hive they also feed and clean the Queen, look after the young grubs, guard the hive entrance against robber bees, and in hot weather fan with their wings to keep the inside of the hive cool. After about a fortnight spent on these tasks they start work outside the hive, collecting *water, propolis, nectar,* and *pollen*. Younger Worker bees then take over the inside work.

Propolis *is* a sticky substance gathered from the limbs and buds of trees. It is used by the bees as a kind of cement with which to fill up cracks in the hive and make the hive weather proof.

Nectar is a sweet liquid secreted by flowers. The foraging worker bee takes it up with her long tongue and passes it into the honeysac that she has inside her body. If it is to be food for the bee herself, she allows it to pass from the honeysac into her stomach, where it is digested. If, on the other hand, it is to be stored, she carries it back to the hive in her honeysac and disgorges it with the assistance of other bees into the cells (comb). During the flight home the nectar in the honeysac undergoes certain changes, and some of the water contained in the nectar is removed. The taking up of nectar by the bees after it has been deposited in the comb continues the drying process. The changes that occur within the honeysac together with the further drying result in the nectar being converted from nectar to honey.

Pollen is also gathered from the flowers and is carried to the hive in the pollen-baskets that each worker has on her third pair of legs. It is mixed with honey to form bee bread, which is kept in reserve so that there is always a supply of pollen in the hive throughout the year. Worker bees modify a honey and pollen mix in their hypopharyngneal gland to feed their larvae. A strong stock will eat very large quantities of pollen in a year.

Pollen if not needed for immediate use it is stored in the cells, and combined with a little honey. Cells containing pollen are not normally capped with wax. The addition of honey to the pollen ensures that it will keep.

Each variety of flower has a pollen of a different colour, and the bees can be seen carrying black pollen from poppies, pale green from the apple, golden from the sainfoin, and many other different coloured pollens.

Another of the workers' duties is to produce wax. This is done in special wax pockets on the underside of the *abdomen* (stomach). This wax is used for building and sealing the cells.

The underside of a worker bee abdomen with wax scales.

Wax is produced only by the worker bees. The wax is exuded from their wax glands in a similar way to how we sweat. The worker bees cluster tightly and the bees in the center of the cluster are heated, this heating induces the production of wax.

Worker bee sting. The worker has a sting shaped like a multi-barbed arrow. She only uses it when she is obliged to; the barbs prevent her pulling it out again. She tries to do so, but in her struggles part of her body is torn away and she dies.

The worker bees are the smallest in the hive and like the Queen they are female, but unlike her they do not lay eggs, except in special circumstances.

The Queen. In each hive there is only one Queen, she is the mother of the colony and her only duty is to lay eggs. The Workers feed her and keep her clean.

Her body is longer than that of the Worker. Her sting is curved and is used against other Queens, rarely against human beings. Her wings are short, as she only flies on special occasions. She has no pollen-baskets on her legs and no wax-making pockets on her body.

So you want to be a beekeeper? - *An Introduction to Beekeeping*

The number of eggs that she lays depends upon the amount of food that she is given by the workers. In the winter she is given very little, but as spring draws near she is given more, until, in the summer, she is given enough to allow her to lay a large number of eggs. She may lay as many as 3,000 eggs in a day, placing each one in a *cell* in the brood-combs.

The Queen lives much longer than the other bees. She may live to be four or five years old, but by that time she has become too old to lay worker eggs, and she is then generally killed by the bees in the hive and superseded with a new queen.

The Drones. The Drones are the largest bees in the colony. They are males and their primary duty is to fertilize virgin Queens. They cannot collect nectar, they do no work inside the hive and have to be fed by the workers. As they have no sting, they cannot even act as guards. When they fly they make a loud buzzing noise, like a very big bluebottle and quite unlike the noise the workers and the Queen make. In the autumn they are turned out of the hive to die.

Single bees egg magnified several hundred times.
Actual size approximately 1.5mm long

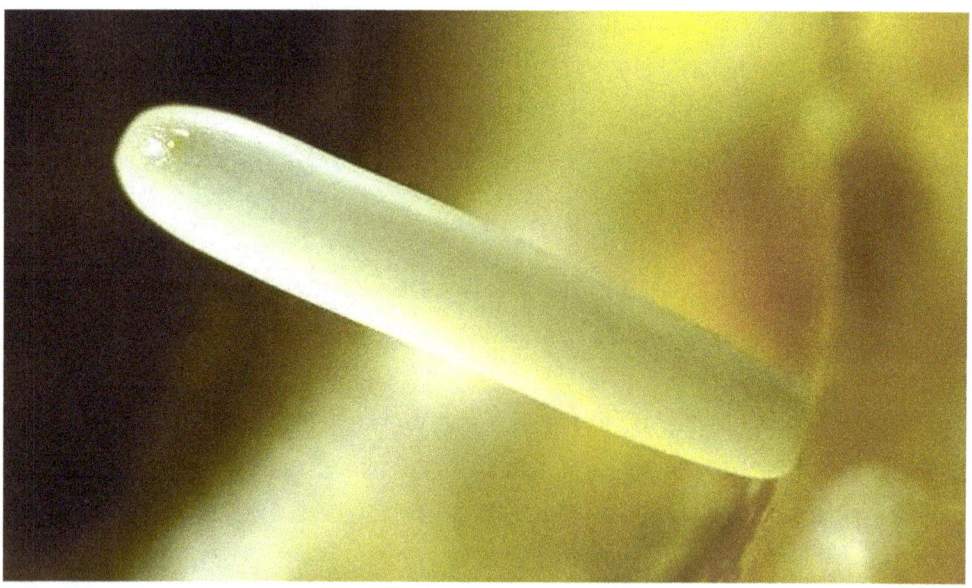

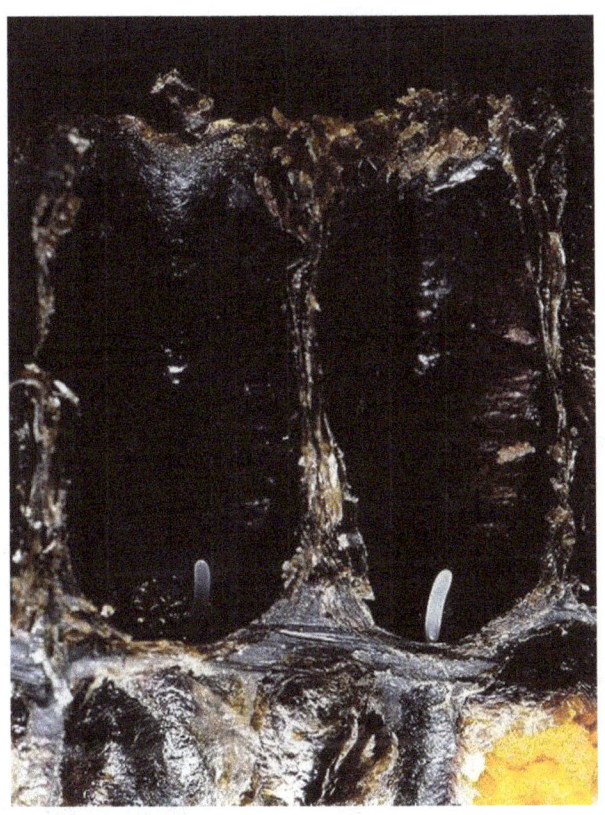

Two cells cut to reveal eggs as layed by the queen
Note they are almost erect, in the first three days the egg wilts from erect to horizontal. Observation of the angle of wilt gives an indication of the age of the egg.

From Egg to Bee. The Queen lays two kinds of eggs. One kind has been fertilized, and these eggs produce larvae that can be made to develop into either workers or Queens, according to the sort of food fed to them. The other kind has not been fertilized, and the eggs produce larvae that always develop into Drones (that is, into male bees).

When the egg is first laid it is very small and white and is stuck vertically to the bottom of the cell. On the second day it leans over, and on the third it lies at the bottom of the cell. If the bee keeper sees eggs in the cells he knows that the Queen is in the hive, and therefore he need not spend time looking for her.

A larva that is being fed so as to become a Queen, needs more room in which to grow than one that is to become a worker. It lives therefore, in a large cell, shaped like an acorn, in the brood-chamber, hanging downwards from the comb. Three days after being laid, the egg hatches into a larva, which is then fed for about five days on a very rich food made by the workers.

A supply of this food, called *Royal jelly,* is then put inside the cell with the grub, and the cell closed *(capped)* with a covering of wax and pollen.

During the next seven days very big changes take place in the little creature shut up in the capped cell. The larva spins a cocoon round itself, turns into a pupa, and then, on the sixteenth day from the laying of the egg, bites through the capping and crawls out on to the comb, a perfect young virgin Queen.

The cell in which a worker is to develop is smaller than a Queen cell. The egg takes three days to hatch into a larva. This is then fed for three days on a rich food and for another three days on a less rich food (a mixture of honey and pollen). Like the Queen cell the worker cell is capped on the ninth day. After that the larva spins a cocoon, develops into a pupa, and bites its way out on the twenty-second day. For the next fourteen days she works inside the hive, and after that starts her duties outside the hive.

The egg that produces the drone is an unfertilized one, but in other respects is like that of the Queen and worker. It is laid in a drone cell, which is larger than the worker cell, as the drone is a larger insect than the worker. The Queen cells are, however, the biggest of all. The larva of the drone is fed in much the same way as the larva of the worker, but the capping that is put on the cell on the ninth day is larger and dome shaped. The Drone takes even longer to develop than the worker, and does not bite through the capping until the twenty-fifth day after the egg was laid.

So you want to be a beekeeper? - *An Introduction to Beekeeping*

The rates of development of the Queen, worker, and drone can be arranged for comparison in this way;-

The time table from egg to perfect bee (imago) in days

Stage		Queen	Worker	Drone
Egg (from laying to hatching)		3	3	3
Larva	before sealing	5	5	6
	Sealed and spinning	1 } = 8	2 } = 10	3 } = 13
	Quiescent	2	3	4
Pupa	transformation to sylph	1 } = 4	1 } = 8	1 } = 8
	Quiescent	3	7	7
Total time from laying of egg to emergence of imago		15 days	21 days	24 days
Period spent in hive before flying		5 days	14 days	13 days
Total time from egg laying to leaving the hive		20 days	35 days	37 days
From the time when the queen normally leaves the hive (20 days from egg laying) she has approximately 20 days in which to get mated by drones, after which she is too old.				
The 14 days that worker bees spend in the hive after emerging are spent as nurse bees and they have specific duties including feeding the queen				
The 13 days that drones are confined to the hive after emerging are spent in maturing. Drones do not appear to have "in hive" duties and are for the insemination of queens only.				

The developmental times shown above can shorten or extend under extreme weather conditions of heat or cold.

Larvae a few days after hatching from the egg.

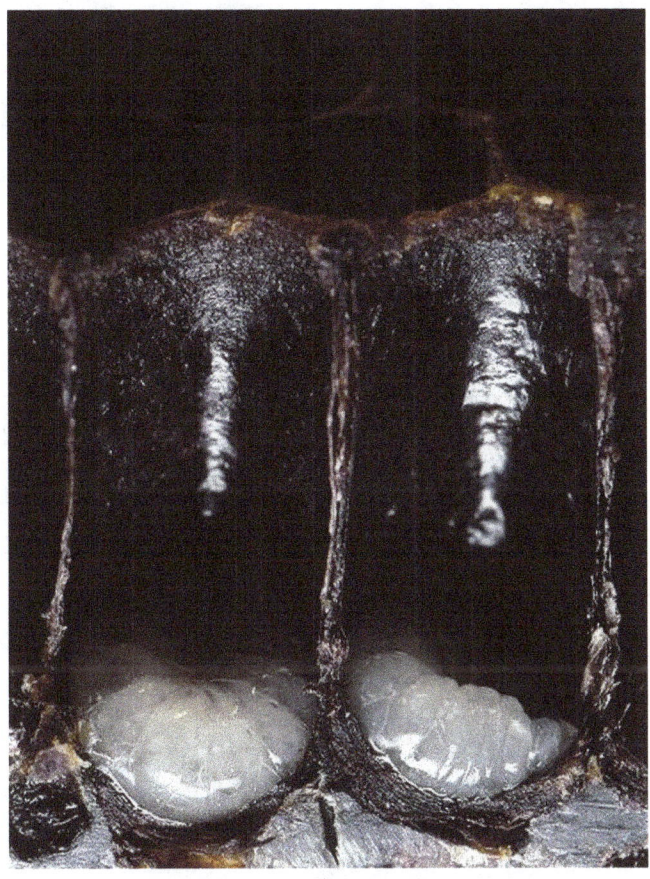

Note that the larvae are pure white and glistening. If the larvae are not as above ie dull or dis-coloured then this can be an indication of a health problem within the hive. Larvae that appear to be off white with an internal lump that resembles a yellowish / slightly curved grain of rice, is indicative that the larvae may be infected with European foul brood.

Uncapped cells showing larvae at the sylph stage, a point in time when they are extended along the full length of the cell immediately prior to cocoon spinning and the commencement of final pupation.

Four cells cut to expose pupae in the later stages of pupation from larvae to imago.

Races of Hive Bees. There are various races of bees, just as there are many races of human beings. There is the English Brown Bee (sometimes called the English Black Bee), the Yellow Italian Bee, the Dark Dutch Bee, the Grey Caucasian Bee, and many others. The best bees to keep should always be native British bees, The bees in the photographs in this book are hybrids or crossbreds. Many bees in this country are hybrids, for the native English Brown Bees died during the epidemic of Isle of Wight disease, (acarine) and fresh stocks of other races were imported from abroad. The true British bee in reality now no longer exists and what most beekeepers refer to as British bees are in fact cross-breeds, or mongrels.

Bee collecting pollen from pussy willow

The pollen that bees gather from flowers is dry and powdery. Bees as well as gathering pollen, are also collecting nectar and they add a little nectar to the pollen as they collect it enabling them to create the nodules (loads) of pollen that are held in place by special hairs on the bees hind legs. When observing the pollen loads they are always of a single colour. This is because when foraging, the bees visit only a single type of flower on a single foraging flight. This selectivity is continued to a significant extent in the hive and when examining the comb the cells containing the stored pollen appear as patches of little paint pots containing nearly every colour of the rainbow.

THE HOME OF THE BEE

Natural Home. Having filled themselves with honey, before leaving their previous home (hive) in the form of a swarm, bees will occupy any reasonable empty space. The most natural being a hollow in a tree-trunk. After only a few hours after arriving in their new home the bees start to build comb in exactly the same way as they would in a skep. See previous photograph of an inverted skep. There they would expand the comb and colony to the full extent of the space available.

The Modern Hive. Until the coming of the wooden hive with movable combs very little was known of the work inside the colony, and even to-day there is much more to be learned. The modern beekeeper tries as far as possible, to copy the natural home of the bee. The outer casing of the hive takes the place of the tree-trunk. . The beekeeper supplies his bees with movable wooden frames in which the bees can build their combs. In each frame is fitted a sheet of wax, which has pieces of wire sunk in it to make it strong. This sheet of wax serves as a foundation for the comb and saves the bees work. The frame / combs can be taken out and looked at. On each frame are two spacers, one at each end, to keep the frame / combs as far apart as they would be in a tree. The terms frame and comb, as used by beekeepers, may be sometimes a little muddling. A *frame*, however, merely means a wooden frame with wax foundation before the bees have built any cells upon it. After the cells have been built it is called a comb.

Brood-Chamber. The bottom portion of the hive, called the brood-chamber, is used by the bees as their home throughout the year. In the summer the combs are used as a nursery in which the Queen lays eggs and the workers rear the bees. In the winter the brood chamber also serves as a cupboard in which food is stored together with brood.

Honey-Chambers or " Supers." In the summer other chambers are placed above the brood-chamber. These contain either shallow frames or small square sections, and on them the bees build cells and store the honey that they do not at that moment need for themselves, ie the surplus. This is the honey that we take from them.

Queen Excluder. To prevent the Queen laying eggs in the honey-chambers, a Queen excluder is placed on top of the brood-chamber. This can be either a thin metal sheet which is slotted, or in the form of stiff wires across a frame spaced such that workers bees can get through but not the Queen.

Queen cup

Observation of the queen cups gives a good indication of the preparations to swarm. Sometimes eggs can be seen in queen cups, this is not necessarily a good or true indication. It is when a small larva can be seen swimming in a sea of white food (royal jelly) that you know fully that swarming is imminent. Within the hive there can be anything from one or two to twenty plus queen cups with larva and royal jelly (charged).

The queen cells shown here are in the transitional stage from queen cups to queen cells proper and are fully charged. Notice the volume of food (royal jelly) that the larvae appear to be swimming in.

There are basically three things that determine as to whether a normal fertilized egg will become a worker or a queen.

Firstly, the size of the cell. A worker cell is less than 5mm across whereas a queen cell is approximately 9mm across.

Secondly, the type of food. The food that the larva that is to become queen is fed is a special diet ie royal jelly.

Thirdly, the quantity of food. It can be clearly seen that the larvae are swimming in a sea of food and are almost force-fed, whereas a worker is fed on demand and when examining worker larvae in their cells no food can be seen.

Queen cells

Both of the queen cells shown have been produced in the normal process of swarming. They are large and well placed on the comb. The first queen cell is on a section of comb containing mostly worker brood. The second is on a section of comb containing both nectar and capped honey.

So you want to be a beekeeper? - *An Introduction to Beekeeping*

The queen cells shown here are emergency cells; produced as a result of something besetting the queen. The bees have unexpectedly found themselves without a queen and have gone full steam ahead to replace her. Caution is thrown to the wind in the bee's desperation. Any queen will do even a small underfed one to tide them over.

LIFE INSIDE THE HIVE

Winter

In winter most insects hibernate in one form or another. Honey bees however, are always active to some degree. On a mild day, even at Christmas, they will leave their cluster on the combs in the brood-chamber and leave their hive for a short flight.

Spring

All creatures feel the call of spring. When the warmer weather comes, the Queen wasps and Bumble Bees may be seen searching the hedgerows for suitable nesting-places. The honey bees, however, already have their home, and their activities start with the Queen laying eggs. An early sign of this laying of eggs is the sight of the bees returning home with large loads of pollen in their-pollen-baskets. With the coming of the spring flowers egg-laying increases, until by May the hive is overflowing with bees.

In the fruit-growing parts of the country many hives of bees are kept in the orchards. Now we have oil seed rape early in the year, so in order to obtain as many bees as possible to work amongst the blossom the bee keeper feeds the bees with sugar syrup. This feeding has to be started six or seven weeks before the blossom comes out, because it takes about five weeks from the time the egg is laid to the time the bee starts her work outside the hive. This syrup is made by dissolving 2 lb. of white sugar in one pint of boiling water, and is fed to the bees by means of a feeder.

As soon as the weather becomes warm enough the beekeeper opens the hives for the inside to be cleaned. This is called spring cleaning. It is very important because it gives the first opportunity since the previous autumn for the bees to be examined.

Drinking-water. With many grubs in the hive quite a lot of drinking-water is wanted, and if there is no running water near-by a drinking-fountain must be provided. At the height of the breeding season a strong stock will drink about one pint of water a day. A saucer of water, with several pebbles in it to save the bees getting drowned, or an upturned jam jar filled with water and standing in a saucer will do. It is important to see that water is always there and is always fresh. The best water to use is clean rain water, to which is added enough common salt to make a 0.1 per cent solution, i.e. one ounce of salt to 64 gallons of water.

Summer. As the spring advances and flowers bloom in plenty, the bees begin to fill the combs with honey. The beekeeper must then provide the special frames in which to store it. These may be either sections or shallow frames.

When the cells are filled and the honey is ripe the bees cover them with a capping of pure wax. When all the combs are completed in this way they are ready to be taken off the hive. The sections give the small square honey combs, while the shallow combs will have their honey extracted.

Removing the Honey. The first thing is to get the bees off the combs containing the honey. They could be brushed or shaken off, but that would take a long time and would make them angry, so a special appliance, called a clearer board, is used. This is a piece of board, which completely covers the top of the brood-chamber. In the middle of it is a hole in which is fitted a small metal gateway, called a bee escape. Inside the bee escape there are two pairs of small springs that allow the bees to pass through one way, but not to return. The super (honey-chamber) is placed on the clearer board, the bees finding themselves separated from the rest of the bees in the hive, descend through this little gateway and cannot get back. The board is usually put on in the evening, and when the beekeeper goes to the hive the next day he finds that the combs containing the honey that he wants to take away have no bees on them.

Inside of a bee escape showing the small springs.

The combs are then taken to a suitable room. This room must be made so that no bees can get in, otherwise, when honey is being extracted from the comb, they will come in thousands to take it back. Alternatively extracting can be done after dark when the bees are not flying. When bees are stealing they get very angry and fight amongst themselves. For this reason honey should never be left about where the bees can get at it.

Extracting

If however shallow combs are being used there is much more work to be done. To get the honey out of the combs the cappings have first to be cut off. This is done with a hot, sharp knife. The combs are then put into a machine called an extractor, which whirls them round and round and flings out the honey. In the bottom of the extractor is a tap, and through this the honey is run into a straining appliance called a ripener. In this is a strainer that takes out all the little pieces of wax that have got into the honey during the extracting. In the honey are also a lot of little air-bubbles, but if it is allowed to stand for a time in the ripener these bubbles rise to the surface and form a sort of scum, which can be skimmed off. The ripener has a tap at the bottom, and through this the honey can be run into jars or storage buckets. The jars are labelled and are then ready to be sold. The empty combs are either put back in the hive to be refilled or cleaned by the bees or packed away until the next season. Putting back the combs saves the bees a lot of work, for then they do not have to build new cells each season. Sections, however, are different, for the honey is sold on the comb, so that a new comb has to be built each time. That is the reason why honey in sections is always more expensive to buy than honey in jars.

The wax cappings that the beekeeper cuts off the shallow combs before extracting the honey need not be wasted. They can be strained through muslin or flannel to recover as much honey as possible and the wax melted and then poured into moulds to set and form hard blocks of beeswax. This beeswax can be sold for making into sheets of wax foundation for frames, or into polish, or church candles. Many beekeepers make their own polish or other products such as soap and hand creams from the wax.

If sections are being used all that need be done is to clean any propolis off the frame and put them into little cases ready to be sold.

Swarming. Spring and early summer-time is the time for swarms as well as the time for honey. When the hive is crowded with bees, and there is plenty of brood and honey, some of the bees- it may be 16,000 or even 40,000 - leave the hive with their Queen to make a new home. Before they leave they make sure that there is a young queen in her cell who can one day become Queen of their old home, that there is plenty of food and brood, and that there are enough bees to look after it all. Then they fill themselves with honey, enough to last them several days, and leave the hive in a swirling, buzzing throng. In the meantime, scout bees are sent out to find a new home, and if they do not come back, the bees form a huge cluster on a bough or in a hedge, or even on a shop-window. They hang together, by their feet, in a huge pear-shaped bunch, and keep very still and quiet. When the scouts return, the bees, if not disturbed before, suddenly break cluster and fly away to their new home, this time flying very quickly. They may stay in the primary cluster for quite a short time, half an hour or so or they may stay a whole day. About half an hour is the average time. Sometimes, if a home is not found, they will build their combs in a hedge. If they do this, however, they rarely survive the cold winter.

Swarm of bees in a fruit tree

To Take a Swarm. A swarm is very valuable to the bee keeper, and he does his best not to lose one. If the bees have clustered on the branch of a tree, he gets a box or a straw skep and holds it under the cluster. Then he gives the branch a shake and the bees fall into the box. He then turns the box and bees upside down on the ground, in the shade of a tree, putting under one side a stone or a piece of stick, so that any of the bees that are flying around may rejoin the swarm. There he leaves them until evening, when he puts them into a hive. To get them to go in he props up the body of the hive, to make a nice large doorway, and places in front of it a wide board with one end resting on the alighting-board of the hive and the other on the ground. Over this he puts a large white sheet, and on to the sheet he shakes the swarm. Bees will nearly always run uphill, and the sloping board induces them to move upwards towards the hive. So up they march, like an army of soldiers into their new home. The sheet is not absolutely necessary, but it gives the bees a nice surface to walk on and makes it easier for the beekeeper to see the Queen. When he has seen the Queen enter the hive he knows that the swarm is safe for the time being. If the sheet is arranged to touch the ground on either side the bees will be unable to get underneath the board.

Of course, bees do not always form a cluster on a branch of a tree. They choose all sorts of queer places, in a hedge, on pea-sticks, against a shop-window, or on a lamp post. If they swarm in a hedge it may be possible to place an inverted box above the cluster. Into this the bees will very probably crawl, for they always tend to move upwards and into the dark. A gentle puff or two from the smoker will hasten them. Bees can generally be swept off a smooth surface like a window, if a brush with soft bristles or a large feather is used. Some beekeepers use their hands instead of a brush.

If the Queen can be found, take her out of the cluster and put her in a matchbox. Make a few holes in the matchbox for ventilation and then place it inside the box that you want the bees to enter. The swarm will nearly always follow, for they can sense where their Queen is and know that they must stay with her. But what of the stock from which this swarm of bees has come?

A few days after the swarm has left, a virgin Queen comes out of her cell. There are many bees left in the hive, and if they decide not to make a second swarm, or cast, with the virgin Queen, she examines each comb and stings to death any of her sister Queens that there may be in the cells. A few days later she leaves the hive for a few short flights to try her wings and survey her surroundings. Then she flies out once more, but this time is followed by the Drones. Drones by whom she is fertilized die very shortly afterwards, but the virgin Queen returns to be Queen of the hive and to start her work of laying eggs. A Queen rarely leaves the hive except on her marriage flight, or to leave with a swarm,

Sometimes a swarm is returned to the hive from which it came, but before returning it the beekeeper examines each comb carefully and destroys all the Queen cells. He also gives the bees more combs in which to store honey, and puts more combs in the brood nest so that the Queen has plenty of cells in which to lay eggs. If this were not done the swarm, for lack of room to expand, would again leave the hive, and might this time be lost.

To prevent Swarming. When a swarm leaves the hive much working time is lost, for the bees do very little work a day or so before swarming. There are many ways in which we can try to prevent swarming. One way is to examine the combs every ten days and destroy all Queen cells, for the bees will rarely swarm if there is no young Queen in a cell to take over the duty of Queen.

Another way is to make what is called a nucleus. Four combs of bees but not the Queen are taken out, three of them containing plenty of eggs and brood, and perhaps a Queen cell, and the other full of honey. These are put into a small hive, called a nucleus box by the side of the original hive, and fed slowly with sugar syrup. Four frames, each containing a sheet of wax foundation, are then placed in the original hive, and more combs are given to the bees in which to store honey. This gives them plenty of room and therefore discourages swarming. In due course a virgin Queen will be born in the nucleus box, either from a Queen cell that was on one of the combs, or reared from one of the eggs.

Later she is fertilized and begins to lay eggs. If left, the nucleus with continued feeding will grow into a full stock.

Uniting. Should it be necessary to unite two stocks of bees, the "newspaper" method is often used. The crown board of one stock is removed and in its place is put a sheet of newspaper in which a few holes have been pierced with a knife.

The other stock is then placed on top of the first, so that in order to pass from one chamber to the other the bees have to chew holes in the paper. By the time they have done this both stocks have the same smell and will join together without fighting. Before placing the second stock on the first every effort should be made to find and remove one of the queens. Failure to do so may result in the queens fighting and the surviving queen being left injured and incapable of egg laying.

Buying Queens. Instead of rearing one's own Queens they may be bought from bee keepers who specialize in Queen rearing. A Queen, with a number of Worker bees to look after her, is sent by post in what is called a Queen travelling cage with two compartments. In one of the end compartments is a lump of fondant and in the other the Queen and her attendant workers. One side of the cage is covered with wire gauze, the end near the fondant, is covered with a lid.

When the cage arrives the lid is removed together with the attendant workers and the cage is put between the tops of the combs of the queenless stock.

Every stock of bees has a different smell, and if a strange-smelling Queen were suddenly introduced into a stock the bees would kill her at once. The bees, however, go for the fondant through the uncovered hole in the cage, and by the time they have eaten it and reached the new bees the Queen has got the smell of the bees of the stock, and is therefore accepted as one of their own.

She joins the bees in the hive by passing through the compartment of the cage that was at first filled with fondant. If the attendant workers are left in the cage they are usually killed.

Autumn. As the summer advances the Queen lays fewer and fewer eggs, and these only in the middle combs of the brood-chamber. The beekeeper has by now removed any supers, and the brood-combs are being filled by the bees with a store of honey for use during the coming winter. The number of flowers for the bees to work, however is gradually getting less, and life in the hive becomes quieter.

Sometimes, in order to provide plenty of food, a full super is left on the hive. The Queen excluder is removed so that the Queen can accompany the other bees when they go up to cluster on the honey-combs. If she was prevented from doing this she would be left behind in the brood-chamber, all alone, and the winter cold would kill her. The excluder is replaced in the early spring and the Queen returned to the brood-chamber.

Killing of the Drones. Another sign of autumn is the turning out of the drones. They are not needed in winter, and if allowed to remain would only eat the precious stores of honey. The workers first starve them, and then, when they have grown weak, they are turned out of the hive. The workers do not sting them, but if any signs of resistance are shown they bite the drones wings and tumble them outside the hive to die.

Feeding, spring. In the spring the queen must be persuaded to increase and build up egg laying. She will only do this if there is food coming into the hive, therefore slow stimulation feeding with a light syrup, may be required.

Feeding, Autumn. Autumn is a very important time for the beekeeper, for the number of stocks that will survive the winter very largely depends upon the work he does now. He must see that each stock has enough honey, stored and capped to last until the flowers come again in the spring, and that there are enough bees to maintain warmth during the long, cold months. Although bees, like all insects, are called cold-blooded, their little bodies produce heat, and this comes from the food that they eat. The greater the number of bees in a stock the greater the amount of heat produced. One bee alone in a hive would certainly die, whereas 10,000 bees would keep each other warm enough for most of them to survive the winter.

Rapid Feeding. If the beekeeper thinks that there is not enough food in an individual hive the beekeeper feeds sugar syrup to the stock so that the reserves of food for the winter can be made up. Each stock needs at least 35 lb. of honey / syrup as a reserve, and if there is not that quantity in the hive, then the beekeeper must feed more sugar syrup. The bees have to put this into the cells and cap it over, just as they do the honey. As it may have to stay in the cells for many weeks it must not contain too much water, otherwise it may ferment. He therefore makes this syrup thicker by dissolving 2 lb. of white sugar in 1 pint of water. This is put into a large a feeder which is put above the brood-chamber, and fed as quickly as the bees will store it. It must be remembered that even thick syrup has much too much water in it, and the bees have to remove a great deal of the water before they can cap and store it. Because bees can only remove that water when the humidity of the air is low, feeding once the autumn mists have arrived it is too late to feed. Syrup taken down too late in the autumn will not be dried, and consequently not capped, and will ferment. Fermenting or fermented syrup will kill the bees.

Feed with lid removed.

Feeder with lid and safety cover removed. The safety cover prevents the bees getting into the syrup and drowning. Bees crawl up the centre tube over the top and down to the syrup surface and sup.

Winter. With plenty of food in the hive, and a strong force of bees to create warmth, all is ready for winter. If quilts are used rather than crown boards some small pieces of wood are placed on top of the frames in the brood chamber. These make a small space between the tops of the combs and the quilt, and through this space the bees will pass as they move slowly over the combs If they had to pass down to the bottom of the combs they might be killed by the cold. The top of the hive is the warmest part.

On top of the crown board or quilt a little packing, such as a piece of clean sacking is placed to help keep the hive warm. A fastening is then put round the outside of the hive to prevent the roof' being blown off in the winter gales. Finally, under the back legs of each hive, small blocks of wood are placed so that the hive is tilted forward slightly. This helps to prevent the inside from becoming damp.

What the Bees do in Winter. So much for the bee keeper's work, but what of the bees? As the weather becomes colder they cluster together on the combs, eat the honey that is in the cells, and so keep up the temperature within the hive. The food is not eaten quickly, and as the cells are emptied the cluster moves on slowly from comb to comb. The wax cappings which they take off the cells are dropped on to the floor of the hive, to be removed by the beekeeper when he examines the varoa board or cleans the hive the following spring.

Fondant. All through the winter the beekeeper has an eye upon his hives. By gently lifting the back of each hive (hefting) he can tell roughly the amount of food left inside. In a mild winter more may be eaten, as the bees are more active. If he thinks that there is a shortage of food he gives each stock a cake fondant, and this is replaced by another as it is consumed.

Production of Honey. The amount of honey gathered by a stock of bees in a season varies very much with the district, the weather, and the number of bees in the stock. It may be as low as 50 lbs. or as much as 400 lb. This includes not only the surplus honey that the beekeeper removes, but also the amount eaten by the bees themselves. The amount eaten by a colony must be added to the amount we remove if we want to calculate the total amount collected by the bees.

For the whole of the country the average amount of surplus honey removed by beekeepers is probably 25 to 30 lb. (1 super) per hive (that is, per colony). A record amount removed from one hive was 312 lb.

Bees, like farmers, are most successful on a suitable soil in a good season. The best type of soil is one with a chalky subsoil, for on this the flowers seem to make large supplies of nectar of good quality. The best season is one with sunny days, warm nights, gentle rains, and very little wind.

Honey varies in colour from black to almost white. The colour depends partly upon the flower from which the bees are collecting nectar, and partly upon the soil in which the flowers grow. For example, beans growing on a chalky soil give nectar that results in a pale-coloured honey, but if the beans grow on a clay soil the honey is much darker.

Ling heather honey is peculiar in being so jelly-like that an extractor cannot deal with it. It has to be pressed out of the combs by means of a honey press. There are fruit presses available that are not designed to press honey but work very well.

THE USEFULNESS OF THE BEE

Each type of honey has its own aroma (smell) and its own flavour, reminding one of the flower from which it was gathered. Honey made from the nectar of fruit blossoms has a faint though definite flavour of plums. Hawthorn honey has a very rich, nutty flavour. Heather honey tastes rich, too, but with a faint, bitter background of taste. Try and identify by the aroma, taste, and colour the flower from which most of the honey in a sample was taken. As the aroma is very delicate and quickly lost, the lid of a honey jar should not be left off longer than is necessary.

Most samples of honey are produced from a mixture of many flowers, although there will usually be more of one kind than another. If extracted honey is kept for some time it sets and becomes granular, this is called granulation. It happens to all kinds of good honey except heather honey. The honey can be made liquid again by standing the jar in warm water.

Once honey has granulated and been melted again it does not granulate again in quite the same way. Honey from the flowers of the Brassica tribe (oil seed rape) is an exception to this rule, for this is honey that granulates very quickly.

Pollination. The gathering of nectar is only one of the ways in which the bee is useful. Indeed, her chief work is the carrying of pollen from flower to flower (see page 27), and thus making possible the production of fruit and seed. Pollen is a fertilizing dust that has to be taken from the male portion of the flower, the stamen, to the female portion, the stigma, and, as the flowers themselves cannot move about, it has to be taken by what are called *agents*. The principal agents are the wind, water, and insects.

In some cases gardeners pollinate flowers by hand, dusting the stigma with a rabbit's tail brush dipped in pollen. Amongst the insects, the most important pollinators are the bees. In order to attract the insects the flower makes nectar. This is usually found deep down in the flower. Along comes the bee to collect it and make it into honey, and in pushing her way into the flower she brushes against the stamens and some of the pollen becomes entangled in the hair on her body. Some of it she cleans off and puts in her pollen-baskets, but some is left on her body and, when she visits the next flower, rubs off on to the sticky stigma. In this way, the flowers are pollinated and crops of fruit and seed are made possible. Some flowers, like the plum, for example, have only one stigma. Others have more. The apple has five, and each one of these must receive this pollen if a perfectly rounded apple is to be formed. You will often have seen an apple that is lop-sided. If you cut it in two, cutting right through the seed-box, or core, and look at the pips, you will find that those on the rounded side are properly shaped, whilst those on the deformed side are undeveloped. This is because one stigma at least was not pollinated.

Farmers who grow crops for seed are helped in the same way by the bees. Such crops include sainfoin, lucerne, mustard and white clover, a plant that is very valuable in the pastures. With bees to pollinate the flowers these plants will set a much heavier crop of seed.

Use of Honey in Olden Times. For many hundreds of years men have made use of the bees. Before the days of the sugar beet and the sugar-cane, honey was the principal source of sweetness. Mead used to be a very important drink, and this is made from honey. The wax, too, was much used for lighting, and even today makes church candles. At one time it was the custom, at a wedding, to make the bride a present of a skep of bees, so that she might have honey and wax for use in the home.

ENEMIES AND DISEASES OF THE BEE

The Enemies of the Bee. Snow is a danger, for it may drift inside the hive.

Bees have their enemies. Mice, knowing it is nice and warm inside the hive, creep in and make a cosy nest in the hive. Unfortunately they gnaw the combs, which disturbs the bees and causes them to eat more honey than they would if they were allowed to remain undisturbed. If the mice are not caught or turned out the bees may die of starvation. To stop mice getting into the hive the entrance should be made as narrow as possible with the use of a mouse guard.

Blue tits disturb the bees in quite a different way. When they are hungry in the winter these cunning little birds sit on the alighting board of the hive and tap with their beaks.

The poor bee that comes to the door to see what the noise is, never returns, for the blue tit has her. The blue tit, however, is a useful bird in other ways, and instead of destroying him the beekeeper hangs a piece of fat a short distance from the hive to provide food for him. If well fed, he will not bother the bees. Another enemy is the wax moth. The female moth enters the hive and lays her eggs in the combs. These eggs hatch into caterpillars that eat their way through the comb, leaving behind them trails of silky web. This spoils the comb. The way to prevent it is to have strong clean stocks of bees and to keep the quilts if used or other coverings clean.

Other enemies of the bees are woodpeckers, wasps and hornets. There are also lesser enemies such as small birds that take insects of all types in flight.

Diseases of Bees. Like all living creatures bees have their diseases. Fortunately for us these do not affect the honey, nor can they be caught by human beings or other animals.

Acarine. formerly called Isle of Wight Disease. This is not really a disease in the ordinary sense of the word, it is caused by tiny mites that invades and lives in the breathing-tubes (tracheas) of the adult bees, blocking these tubes and sucking the blood of the bee. When suffering from this disease bees cannot fly. They will crawl about on the ground near the hive and climb to the top of stalks of grass in an attempt to get off the ground. They hold their wings in queer positions as though dislocated (called K wing). A number of years ago there was a terrible plague of this disease and most of the bees in this country died. The position was so serious that the Government helped to re-stock the apiaries with bees imported from Holland.

Dysentery. (Nosema) This usually occurs in early spring and can be the result of bad management. A damp or draughty hive, disturbance of the hive during the winter, or bad or unripe food may cause it. Sometimes it is a symptom of more serious trouble, such as acarine. It can be identified in its advanced stages by the bees defecating in or on the hive; the deposits that the bee leaves are in the form of small vivid yellow stripes. Quite often when the signs become visible it is too late but transferring the bees to a clean, dry hive, and giving them good food might help them to recover.
Nosema can be detected by microscopically examining a sample of bees. If nosema spores are found in significant numbers the colony can be treated with a suitable antibiotic.

Varroa. Varroa is a small mite (much larger than acarine) that is parasitic on bees. The adult varroa lays its eggs in the cells of developing grubs and emerges with the bee when it chews its way out of its cell. Its primary host is the drone grub it is also found in the worker bee cell. Varroa does not however appear to breed in queen cells. There are a number of miteicides for the control of varroa and advise about their use, can be obtained either from your local association or a good supplier of bee equipment.

Foul Brood. There are two type of foul brood; American and European. They affect the brood and not the adult bees and in advanced stages of infections are noticeable by their very unpleasant smell, hence foul brood. Because of their seriousness, they have to be reported the regional bee inspector. It is best to read more about these and other diseases in leaflets supplied by DEFRA

LIST OF WORDS USED IN BEE KEEPING

Apiary A place where one keeps a hive, or hives, of bees.

Apidea Miniature hive used in the breeding of queen bees

Bee Bread Pollen and honey mixed by the bees for feeding larvae.

Brood The eggs, larvae, and nymphs in the brood-chamber.

Brood Comb Used in the brood-chamber for rearing bees in the summer and storing food for winter.

Cast A second or subsequent swarm, containing a virgin Queen.

Cluster A swarm that has settled in a tree, bush or elsewhere, or bees forming a ball like structure within the hive during winter.

Comb See Frame.

Colony See Stock.

Eke A spacer, the same size as a super or brood box used to provide space for feeding etc

Fanning Action of bees to move air for cooling or sending out pheromones.

Frame A wooden surround in which is placed a sheet of foundation. On the wax the bees build their cells, and it is then called a *comb*.

Fondant Sugar paste with very low water content.

Foundation A thin sheet of wax with embossing on both faces in the form of hexagons.

Granulation The change of honey from the liquid to the solid or semi-solid state.

Hive Tool Tool used to prise parts of the hive open.

Honey Dew Honey processed by bees from the excretions of aphids.

Honey Sac A small bag inside the body of the bee in which she carries nectar to be turned into honey. Water can also be carried in this sac.

Hypopharyngneal gland. Gland in the worker bee's head that modifies food to produce "jelly" for larvae

Imago. The final fully developed insect (the emerging bee)

Larva The stage in the life between egg and pupa of an insect. The larva of a butterfly or moth is called a caterpillar. The larva of a bee is sometimes called a grub.

Nectar A sweet liquid found in the flowers and collected by the bees and turned into honey.

Nasonov gland. Gland on the end of a worker bees abdomen that emits pheromones.

Nucleus A small stock of bees on combs.

Nymph Used by bee keepers to mean the pupa.

Princess A young virgin (unmated) Queen.

Propolis A sticky substance gathered by the bees from shrubs and trees, used as a filler and cement in the hive.

Pupa The stage in the life of the bee that follows the larval stage and comes immediately before the adult stage.

Quilt A linen or sailcloth cover that is placed over the combs.

Royal jelly A rich food made by the Worker bees and fed to Queen larvae.

Section Small square or circular box fitted with a thin sheet of foundation, and when filled with comb and honey is sold as created by the bees.

Skep The old-fashioned straw hive. Now primarily used to collect swarms.

Stock or *Colony.* An organized community of bees. ie queen, drones and workers within a hive.

Swarm A large number of workers, with some drones, and the queen, that leave the hive to make a new colony elsewhere.

Varroa Floor The bottom section of a hive incorporating the hive entrance, mesh screen and removable panel for the counting of varroa.

Wax glands. Glands on the underside of a worker honey bee that produces wax. (in the form of minute flakes)

This diagram is not fully annotated, add other information as required

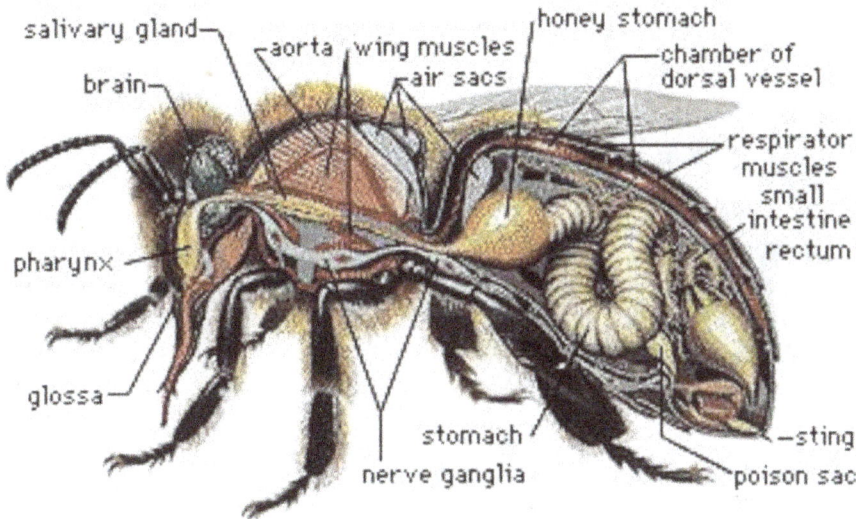

Children from a local school, checking a hive for honey.

Note the boy examining the comb is not wearing gloves.

All beekeepers should develop the confidence from the outset of their beekeeping career, not to need gloves for protection against stings. However thin surgical gloves are very good in preventing hands and fingers from being gummed up with propolis but they offer no protection against stings.

Using thin surgical gloves does not detract from the sensitivity required when handling bees. The use of heavy leather gloves will automatically mean that your actions will be difficult and therefore clumsy and the bees will re-act in a more defensive and aggressive way.

Author with children, observing bees in top bar hives.

The hives were located inside a redundant mobile site office that had been converted as a dedicated experimental bee house.

CONFIDENCE CHECK LIST
ARE YOU HAPPY THAT YOU KNOW OR CAN:-

Light a smoker.
Identify all the main components of a standard hive.
Plain floor, Varroa mesh floor, Brood box,
Super, Crown board, ect
Different types of hive; Commercial types National, WBC, Top Bar and skep.
Use, and types of hive tools.
Difference between worker and drone comb.
Construction and assembly of frames.
Use of spacers and self spacing frames.
Handle frame of bees without gloves
Coped with and removed, first sting
Identify a worker honey bee.
Identify a drone honey bee.
Identify a Queen bee.
Identify honey. (capped)
Identify nectar in the comb.
Identify stored pollen.
Identify worker eggs.
Identify worker larvae.
Identify drone brood.
Identify queen cups.
Identify charged queen cells.
Identify capped queen cells.
Know the difference between capped worker brood and capped honey.
Identify some common pollens using colour chart.
Identify varroa.
Use a smoker.
Use manipulation cloths.

So you want to be a beekeeper? - *An Introduction to Beekeeping*

Remove frame of bees from hive without assistance.
Open, examine and close a hive without assistance.

ACKNOWLEDGEMENT

Additionally to the author's photographs, some are taken from non-copyrighted sources that are already in the public domain.

Front cover picture.
Honey bee sucking nectar from a borage flower

© J R Slade
Member of the
BBKA Holsworthy Branch

www.ingramcontent.com/pod-product-compliance
Lightning Source LLC
Chambersburg PA
CBHW081508040426
42446CB00017B/3436